Manufacturing in Space

by Patrick H. Stakem

(c) 2017, 2022

Number 5 in the Space series.

Table of Contents

Introduction..4
Author..5
Manufacturing in Space..7
    Why in space?...7
What manufacturing processes would benefit?............................8
    Metallurgy..8
    Thin films...8
    Assembly in Orbit..10
    Maintenance in Space...10
    Orbital debris, and dealing with Zombie-sats........................10
Platforms..11
Space Environment..13
    Zero G issues 14
    Vacuum..14
    Thermal environment 15
    Orbital Debris..15
    Mechanical and Structural Issues..16
    Spacecraft Charging..17
    Radiation Environment and Effects.......................................18
    Commercial Space Stations...21
    Lunar base...21
    Asteroids, and asteroid mining..23
    Mars Base..27
The Players..28
    Commercial Spaceflight Federation......................................29
    Axion Space..30
    Orbital Technologies Commercial Space Station.................31
    Bigelow...31
    Excalibur-Almaz...32
    Space Island Group...33
    Secure World Foundation..34
    Goldman Sachs...34
Wrap-up..34
Bibliography...36
Resources...40

Glossary of terms..44
If you enjoyed this book, you might also be interested in some of these..51

Introduction

"I'll make a prediction right now. The first trillonaire will be made in space."

Ted Cruz, Senator from Texas.

This book covers the topic of Manufacturing in Space, which is not that far away, and has actually been done on a small scale for many years. With permanent manufacturing facilities in space, near to lunar or asteroid resources, we will be able to fabricate facilities from local material, and extract rocket fuel. All of this can replace what we now need very large rockets up from Earth's "gravity well." We can build the next generation stations and spacecraft in situ, in orbit. There are some major advantages to this, as fewer parts need to be lifted up to orbit. Spin-off company's, providing logistics services, will be necessary. Space will be evolving as a frontier outpost. We have experience with those. But, space is a harsh environment, harsher than the Klondike during the gold rush. Yet, the gold rush happened.

Viewed in an economic sense, manufacturing "space stuff" in space makes sense. There is a large initial investment, but the reduced costs, particularly of things that are intended to stay in space, will more than balance this out.

Manufacturing of any kind, any where, involves the raw materials, a source of power, a source of labor (some may be robotic), and a transportation infrastructure. Following the 19th Century iron manufacturing model, it is better to manufacture near the source of the raw materials, and ship the finished product to the customer, than ship the raw materials. It is expected to be able to separate out usable amounts iron, aluminum, silicon, and oxygen from lunar and asteroid material. In addition, water ice is know to exist at the lunar south pole. Extremely pure silicon wafers would be a valuable down-cargo. These could also be used in a subsequent process in-orbit to produce solar panels.

For space manufacturing, the power part is easy, sunlight 24x7. It is costly to bring materials up from the surface of Earth. The ideal situation is to use lunar or asteroid material. The customers may be on the Earth, or we may be building something that stays in orbit, or moves on to a different destination, such as Mars. It is expensive to boost skilled workers up from the surface, and keep them alive in orbit, but hopefully we can plan for tele-operation from Earth's surface, plus automated processes.

The rules change a bit in Space, and we will see yet another Industrial Revolution. Manufacturing in Space is by all definitions a Paradigm Shift. NASA's Marshall Space Flight Center in Huntsville, Alabama hosts NASA's National Center for Advanced Manufacturing.

There is now a Space Manufacturing Conference, hosted by the Space Studies Institute.

Author

Mr. Patrick H. Stakem has been fascinated by the space program since the Vanguard launches in 1957. He received a Bachelors degree in Electrical Engineering from Carnegie-Mellon University, and Masters Degrees in Physics and Computer Science from the Johns Hopkins University. At Carnegie, he worked with a group of undergraduate students to re-assemble, modify, and operate a surplus missile guidance computer, which was later donated to the Smithsonian. He was brought up in the mainframe era, and was taught to never trust a computer you could lift.

He began his career in Aerospace with Fairchild Industries on the ATS-6 (Applications Technology Satellite-6) program, a communication satellite that developed much of the technology for the TDRSS (Tracking and Data Relay Satellite System). He followed the ATS-6 Program through its operational phase, and worked on other projects at NASA's Goddard Space Flight Center including the Hubble Space Telescope, the International Ultraviolet Explorer (IUE), the Solar Maximum Mission (SMM), some of the Landsat missions, and Shuttle. He was posted to NASA's Jet Propulsion Laboratory for Mars-Jupiter-Saturn (MJS-77), which later became the *Voyager* mission, and is still operating and returning data from outside the solar system at this writing. He initiated and lead the international Flight Linux Project for NASA's Earth Sciences Technology Office. He is the recipient of the Shuttle Program Manager's Commendation Award, and has completed 42 NASA Certification courses. He has two NASA Group Achievement Awards, and the Apollo-Soyuz Test Program Award.

Mr. Stakem has been affiliated with the Whiting School of Engineering of the Johns Hopkins University since

2007, and Capitol Technology University. Mr. Stakem supported the Summer Engineering Bootcamp Projects at Goddard Space Flight Center.

Mr. Stakem can be found on Facebook and LinkedIn. Comments, corrections, suggestions are appreciated.

Manufacturing in Space

Keep in mind, manufacturing in space has been going on for a long time. Different small crystal samples were made in Crewed Capsules, back in the day. As we were able to launch and support crewed space stations, with some lab space, manufacturing experiments became more common.

Why in space?

Two main reasons. One is, the object does not have to be launch-rated, and survive the rigors of launch. Secondly, if its going to be used in space anyway, why not built it there. You still have to launch the raw materials, until we get good at exploiting lunar and asteroid raw materials.

Power is free, once you launch the solar cells, and these are one time that could be built in space. If we need very high temperatures, we use focused sunlight. If we need very cold environments, we shield from direct sunlight. There is no ongoing cost for "hot" or "cold."

We will be in a zero-g environment, which can be either bad or good. We are isolated, so hazardous materials and processes could be done safely, tele-remotely. If the process goes badly, the pollutant is not on our planet.

Manufacturing in space enables creation of new materials, that are impossible to manufacture except in zero-gravity. In 1969, on the Soyuz 6 mission, Russian Cosmonauts conducted welding experiments with aluminum, titanium, and stainless steel.

What manufacturing processes would benefit?

This section discusses some manufacturing processes that could be done in space. The obvious connection to using materials that are already in space is also mentioned.

Metallurgy

One thing about the space environment is, we can take advantage of its unique characteristics to make materials that we can't do on Earth. For example, foamy metals. Using injected gas, bubbles can be formed in molten metal, and they will stay in place as the metals re-solidifies. In addition, in zero gravity, you can make alloys of materials that would not normally work, due to their different densities.

Thin films

Processing silicone, gallium arsenide, and other semiconductor materials into extremely pure, and large semiconductor blanks would be so valuable to the semiconductor industry that materials could be carried to orbit for processing, before other sources are discovered.

Another target market is advanced semiconductors. The

zero gravity manufacturing environment will provide more uniform material with fewer unwanted impurities. Following on this, is the manufacture of MEMS. The *MEMS*, or micro-electro-mechanical system, uses a chip-level integrated circuit technology to provide measurement devices. The advantage is that the sensors are made in processes developed for semiconductor manufacture, and are inexpensive to mass-produce. There are MEMS devices in your cell phone.

Crystal manufacturing is also a good candidate for zero-G operations, as has been shown on the ISS and previous missions. Manufacturing biological materials also provides results that are not possible in a gravity field. This includes organ growing, which cannot be done in a gravity field. This had been demonstrated on the Shuttle.

The Pharmaceutical industry is interested, as the current pneumonia virus has developed an immunity to the 49 possible variations of Earth-penicillin. Zero-G will provide the ability to produce different variations to the penicillin, that the virus may not be able to overcome.

NASA is participating in the development of technologies for in-space manufacturing with the Future Engineers Program. This was done jointly with the American Society of Mechanical Engineers. There were competitions in the Junior (K-12) and Teen (13-18) categories. The first challenge was to design a tool that could be used on the ISS. The winner's design was sent to the ISS for printing, and evaluation. The second competition involved a 3-d printable version of the standard handrail clamp assembly for the ISS, and there were 500 entries. Five winners were selected.

Assembly in Orbit

The ISS and prior space stations were assembled in orbit from modules limited in size by the lift capability of the launch vehicles. This was done with human crews, assisted by the "cranes" - the Shuttle and ISS arms. Tele-robotic assembly has been demonstrated, and robotic maintenance missions are planned.

Maintenance in Space

Almost since the beginning of crewed spaceflight, interior and EVA maintenance has been conducted, both routine and emergency. These lessons-learned have influenced the design of space hardware to be repairable (and this was shown on the successful Hubble Space Telescope repair, among others. Now there is a big effort in robotic in-space repair, since we no longer have the flexibility of the Shuttle for low Earth orbit. Higher orbits, sun-synchronous, polar, have many broken or out-of-fuel satellites. This is an area that is being actively addressed.

Orbital debris, and dealing with Zombie-sats

If we have a good way to capture space junk, from nuts and bolts to dead satellites, we can de-orbit the debris, or possibly process it with solar reflectors, separating the constituent materials. At the moment, all of this orbital debris is seen as a hazard to other spacecraft and the ISS crew. It can also be viewed as a rich collection of raw material and manufactured goods, placed in Earth orbit at great cost. We need a method, probably robotic, of collecting all this material for further processing, and to get it out of the way.

Platforms

This section discuss the various platforms that will support space manufacturing. The earliest capsules had little space to spare, but the Apollo-Soyuz mission, with two docked spacecraft had some experiment space in the docking adapter.

The Skylab mission included an electric furnace, a crystal growth chamber, and an electron beam gun. Of particular interest was the processing of molten metals, seeing the behavior of burning materials, crystal growth, and constructing alloys of materials that are not mixable in a gravity field. Electron beam welding was also demonstrated on Skylab.

The Space Shuttle also conducted manufacturing experiments, with its Spacelab facility. This was a laboratory in the Shuttle bay, accessed from the crew compartment. It flew on 32 missions from 1981 to 2000, and pioneered a series of experiments in zero gravity. There was a habitable module, and exterior pallets. If the habitat module was not required, pallets alone could be flown. There was an available Instrument Pointing System (IPS) that could aim telescopes and cameras.

Many small scale manufacturing in space experiments have been carried out at the International Space Station. The Columbus module was the location for most of these. There was a Fluid Science Laboratory, studying the behavior of liquids in low gravity. The Materials Science Lab Electromagnetic Levitator was used to study the melting and solidification of various materials.

There are two 3D printers on the ISS, made by a California Company, Made-in-Space. NASA is exploring having the capability of printed custom parts in situ, as opposed to carrying spares, as a key for the Mars mission. Made-in-Space sees the orbital facility as an ideal location to manufacture optical fiber. Currently, items are printed for use onboard the station (as opposed to sending them up on a resupply flight, and other items are printed and returned to Earth for testing. A practical device that was printed was a buckle for the onboard exercise equipment, designed by Astronaut Yvonne Cagle.

ESA's Columbus Lab module is part of the ISS. It has The Material Science Laboratory Electromagnetic Levitator (MSL-EML). That allows the study of melting and solidification. There is also a Fluid Science Laboratory. This may be operated onboard, or by scientists on the ground, termed tele-science. It was constructed by Thales Alenia Space, in Turni, Italy. It went up on Space Shuttle Atlantis, and has long exceeded its operational life. It's control center is the German Aerospace Center.

Made-in-Space is designing an advanced 3-D printer with robotic arms. This could fabricate and install structural members. The advantage of this approach is to have objects that won't fit on the launch vehicle, even with folding. The other advantage to in-space manufacturing is that the item won't have to survive launch vibration and acoustics, even when it spends the rest of its life in zero-G. NASA is funding Made-in-Space, and its partners Northrop Grumman, and Oceaneering Space Systems.

The Industrial Space facility was a private space station proposed in the 1980's, to be built with private funding, by Space Industries, Inc. The founder of the company was Max Faget, who had been Chief of Engineering and Operations at NASA. The concept was to have an un-crewed facility, that could receive temporary life support onboard, during a Shuttle visit. The necessary funding was not forthcoming, and the project never got going. President Reagan had requested $700 million from Congress for the project, but the request was declined.

We will expect to see manufacturing labs on the Moon and Mars, providing material from local regolith.

Space Environment

The space environment is hostile and non-forgiving. There is very little gravity, so no convection cooling is possible, leading to potential thermal problems. On the other hand, the environment eliminated sedimetation.

Space is a high radiation environment, being above the shielding provided by our` atmosphere.

Zero G issues

Zero gravity, actually, free-fall, brings with it problems. There is no convection cooling, as that relies on the different densities of warm and cool air. There is no mixing of gases of different densities, so you could find yourself in a big bubble of carbon dioxide, unable to breath. Any little pieces of conductive material will float around and short out critical circuitry at the worst possible time. And then, there are the strange issues.

The Hughes (Boeing) HS 601 series of communications spacecraft suffered a series of failures in 1992-1995 due to relays. In zero gravity, tin "whiskers" grew within the units, causing them to short. The control processors on six spacecraft were effected, with three mission failures because both primary and backup computers failed. This is now a well known materials issue, with recommendations for the proper solder to be used. In 1998, the on-orbit Galaxy IV satellite's main control computer failed due to tin whiskers.

Vacuum

The spacecraft operates in vacuum. Not a perfect vacuum, but fairly close. This implies a few things. Lubricant disappears. All the materials outgas to some extent. All this material can find its way to condense on optical surfaces. On the other hand, space is ultra-clean.

Thermal environment

In space, things are either too hot or too cold. Cooling is by conduction to an outside surface, and then radiation to cold space. This requires heat-generating electronics to have a conductive path to a radiator. That makes board design and chip packaging complex and expensive. You get about 1 watt per square meter of sunlight in low Earth orbit. This will heat up the spacecraft, or you can convert it to electrical power with solar arrays.

Parts can be damaged by excessive heat, both ambient and self-generated. In a condition known as *thermal runaway*, an uncontrolled positive feedback situation is created, where overheating causes the part to further overheat, and fail faster.

There can be a large thermal gradient of hundreds of degrees across the satellite, where one side faces the sun, and the other side faces cold space.

Orbital Debris

There is a huge amount of debris in Earth orbit, including old booster units, failed satellites (Zombie-sats), broken solar panels, nuts and bolts, and a Russian Space Suit. Space is large, but all of this stuff constitutes a hazard to ongoing missions. We are adding to the problem by launching Cubesats in large quantities. Now, all spacecraft need a disposal/de-orbit plan, before they are launched.

A major hazard to the ISS and other spacecraft is the proliferation of Zombie-Sats in orbit, dead or malfunctioning satellites that pose a hazard to other satellites. These need to be re-entered into the atmosphere, or recycled. The ISS will be in the same predicament before too long. Too large to allow re-entry intact, some of it will be re-purposed for other projects, and some will be allowed to reenter. Robotic spacecraft to grab zombie-sats and put them in a re-entry trajectory have been proposed.

Cubesats need to comply with the National Space Policy requirement for space debris, which says that small satellites in LEO should re-enter the atmosphere within 25 years from launch.

At Capitol Technology University in Laurel, Maryland, they were addressing this problem with a directed Cubesat mission named TrapSat. NASA though well of this idea, and agreed to provide a free launch for the student designed and built project. This mission will capture and image micro debris, to characterize that environment.

Mechanical and Structural Issues

In zero gravity, everything floats, whether you want it to or not. Floating conductive particles, bits of solder or bonding wire, can short out circuitry. This is mitigated by conformal coatings, but the perimeter of the semiconductor die is usually maintained at ground potential, and cannot be coated due to the manufacturing process.

The challenges of electronics in space are daunting, but much is now understood about the failure mechanisms, and techniques to address them.

Another issue in vacuum is the cold-welding of metallic materials. This occurs when two pieces of material, without an oxide layer, are pressed together. This is facilitated by having very clean surfaces, and a vacuum environment. This affects moving subsystems such as solar arrays and steerable antennas. Early deployment of mechanism is usually not a problem, but mechanisms that have to move periodically throughout the mission can be problematic.

Spacecraft Charging

Another problem with on-orbit spacecraft is that they are not "grounded." This can be a problem when a potential develops across the structure. Ideally, steps were taken to keep every surface linked, electrically. But, the changing phenomena has been the cause of spacecraft system failures. Where does the charge come from? Mostly, the Sun, in the forms of charged particles. This can cause surface charging, and even internal charging. Above about 90 kilometers in altitude, the spacecraft is in a plasma environment At low Earth orbit, there is a low energy but high density of the plasma. The plasma rotates with the Earth's magnetic field. The density is greater at the equator, and less at the magnetic poles. Generally, electrons with energies from 1-100 keV cause surface charging, and those over 100 keV can penetrate and cause internal charging. As modern electronics is very susceptible to electron damage, proper management of charging is needed at the design level.

Just flying along in orbit causes an electric field around the spacecraft, as any conductor traveling through a magnetic field does. If everything is at the same potential, we're good, but if there's a difference in potential, there can be electrostatic discharge. These discharges lead to electronics damage and failure, and can also cause physical damage to surfaces, due to arcing.

Radiation Environment and Effects

There are two radiation problem areas: cumulative dose, and single event. Operating above the Van Allen belts of particles trapped in Earth's magnetic flux lines, spacecraft are exposed to the full fury of the Universe. Earth's magnetic poles do not align with the rotational poles, so the Van Allen belts dip to around 200 kilometers in the South Atlantic, leaving a region called the South Atlantic Anomaly. The magnetic field lines are good at deflecting charged particles, but mostly useless against electromagnetic radiation and uncharged particles such as neutrons. One trip across the Van Allen belts can ruin a spacecraft's electronics. Some spacecraft turn off sensitive electronics for several minutes every ninety minutes – every pass through the low dipping trapped radiation belts in the South Atlantic.

The Earth and other planets are constantly immersed in the solar wind, a flow of hot plasma emitted by the Sun in all directions, a result of the two-million-degree heat of the Sun's outermost layer, the Corona. The solar wind usually reaches Earth with a velocity around 400 km/s, with a density around 5 ions/cm^3. During magnetic storms on the Sun, flows can be several times faster, and stronger. The Sun has an eleven year cycle of maxima. A solar flare is a large explosion in the Sun's atmosphere that can release as much as 6×10^{25} joules of energy in one event, equal to about one sixth of the Sun's total energy output every second. Solar flares are frequently coincident with sun spots. Solar flares, being releases of large amounts of energy, can trigger Coronal Mass Ejections, and accelerate lighter particles like protons to near the speed of light.

Planets with magnetic fields will trap energetic particles arriving from the Sun into orbiting bands. Earth's are called the Van Allen Belts, after their discoverer. The size of the Van Allen Belts shrink and expand in response to the Solar Wind. The wind is made up of particles, electrons up to 10 Million electron volts (MeV), and protons up to 100 Mev – all ionizing doses. One charged particle can knock thousands of secondary electrons loose from the semiconductor lattice, causing noise, spikes, and current surges. Since memory elements are capacitors, they can be damaged or discharged, essentially changing state.

Galactic Cosmic rays are actually heavy ions, not originating in our solar system. The actual origin is unknown. They carry massive amounts of energy, up into the billions (10^9) of electron volts.

Vacuum tube based technology is essentially immune from radiation effects. The Russians designed (but did not complete) a Venus Rover mission using vacuum tube electronics. The Pioneer Venus spacecraft was launched into Venus orbit in 1978, and returned data until 1992. It did not use a computer, but an attitude controller built from discrete components.

Not that just current electronics are vulnerable. The Great Auroral Exhibition of 1859 interacted with the then-extant telegraph lines acting as antennae, such that batteries were not needed for the telegraph apparatus to operate for hours at a time. Some telegraph systems were set on fire, and operators shocked. The whole show is referred to as the Carrington Event, after amateur British Astronomer Richard Carrington.

Around other planets, the closer we get to the Sun, the bigger the impact of solar generated particles, and the less predictable they are. Auroras have been observed on Venus, in spite of the planet not having an observed magnetic field. The impact of the solar particles becomes less of a problem with the outer planets. Auroras have been observed on Mars, and the magnetic filed of Jupiter, Saturn, and some of the moons cause their "Van Allen belts" to trap large numbers of energetic particles, which cause more problems for spacecraft. Both Jupiter and Saturn have magnetic fields greater than Earth's. Not all planets have a magnetic field, so not all have charged particle belts.

Commercial Space Stations

The Russians proposed an orbital construction yard in 2008 for payloads too heavy to launch directly from Earth. Pre-fabricated sections would be launched to the facility, and integrated. The facility was called OPSEK – orbital piloted assembly and experiment complex. Modules from the ISS, when decommissioned, would be re-used initially for OPSEK. The station could also be a quarantine stop for astronauts returning from Interplanetary missions. The project has not been implemented.

Another Commercial Space Station project was proposed by Orbital Technologies, a Russian Company. The 2010 design was a single module with a usable volume of about 20 cubic meters. Customers were interesting in pursuing protein crystallization and materials research in orbit. To date, this has not flown in space.

Lunar base

A company named Shackleton Energy Company wants to exploit the water ice of the moon. This is to lead to a network of refueling stations for ships with liquid propellant engines, using liquid water and liquid hydrogen. Bringing up the material from the surface of the moon is less costly in fuel, than from Earth. A study will be needed on the relative economics of separating the hydrogen and oxygen on the lunar surface, and transporting the gases, or lifting the water to an orbiting hydrolysis station. The fuel and oxidizer would be available to government or other commercial entities (if you had their loyalty card.....3 points for every 10,000 gallons).

Shackleton is thinking of a fuel processing facility on the lunar surface, near the sources of water ice, with propellant depots in low Earth orbit. Hydrogen peroxide can also be manufactured, and is of value as a fuel.

Two inexpensive options, compared to rocket launches, to get lunar material to orbit are the space elevator, and the mass driver. The space elevator is a concept that would work nicely on the moon. We have the advantage of a lower gravity than Earth, no atmosphere, and it could be built with currently available materials. There is nothing nearby to interfere with its operation. The lunar elevator would span about 50,000 km. The elevator needs a solid tether on the surface, a large mass at the upper end, for a tether, and a very strong cable. A lunar elevator could be tethered to a mass at the L1 Lagrange point, between the moon and the Earth. An elevator on the back side is also feasible. Space elevators have been explored since the 1890's. We now have the technology to construct them. A handy asteroid could be used as the counterweight for the lunar elevator.

A mass driver is an electromagnetic catapult, utilizing a long, linear motor. This works well on the lunar surface, with its reduced gravity compared to Earth, and lack of atmospheric drag. The other advantage is 15 days of sunlight, to operate the driver as well as charge batteries.

In a mass driver, the payload does not contact the launch rail, but is magnetically levitated, and accelerated.

Asteroids, and asteroid mining

Asteroids have been imaged by the New Horizons spacecraft, on its way to Pluto, and by the Cassini spacecraft. The Pioneer-10 spacecraft was sent to study the far reaches of the solar system It passed through the Asteroid belt on its way to Jupiter and Saturn.

A driver in the space environment is the exploration of the asteroids, numbering in the millions. Although there are fewer than 10 planets, and less than 200 moons, there are millions of asteroids, mostly in the inner solar system. The main asteroid belt is between Mars and Jupiter. Each may be unique, and some may provide needed raw materials for Earth's use. There are three main classifications: carbon-rich, stony, and metallic.

The physical composition of asteroids is varied and poorly understood. Ceres appears to be composed of a rocky core covered by an icy mantle, whereas Vesta may have a nickel-iron core. Hygiea appears to have a uniformly primitive composition of carbonaceous chondrite. Many of the smaller asteroids are piles of rubble held together loosely by gravity. Some have moons themselves, or are co-orbiting binary asteroids. The bottom line is, asteroids are diverse.

It has been suggested that asteroids might be used as a source of materials that may be rare or exhausted on earth (asteroid mining) or materials for constructing space habitats or as refuelling stations for missions. Materials that are heavy and expensive to launch from Earth may someday be mined from asteroids and used directly for space manufacturing. Valuable materials such as platinum and gold may be returned to Earth for a profit.

Exploring the known asteroids is a daunting challenge. On the other hand, the asteroids can be a significant source of raw materials for Earth. A conventional survey and exploration approach would take too long. What is needed instead is a multitude of autonomous and flexible nano-spacecraft. The architectural model is a swarm (social insect model) of distributed intelligence.

The asteroids are not uniformly distributed. In the asteroid belt, the Kirkwood gaps are relatively empty spots. This is caused by orbital resonance of the asteroids with Jupiter. Orbiting irregular shaped bodies is challenging, due to the gravity field. This makes station keeping and attitude control a problem.

We won't run out of asteroids in the short term. No one knows how man there are, but it is estimated to be in the millions. They range in size from very small, centimeters across, to hundreds of kilometers.

Some asteroids, lumped by their spectrum into the C-class have a lot of carbon, and a lot of water, bound in hydrides of a clay-like material. The water could become important, as we can use solar energy to break it down into hydrogen and oxygen - rocket fuel, like the Saturn-V used, and this fuel would already be in space.

NASA is currently investigating asteroids in situ, with the OSIRIS-REx mission. (The author worked on this mission) This involves close up studies, and a sample return. The Mission name is derived from Origins, Spectral Interpretations, Resource Identification, Security, and Regolith Explorer. It was launched in September, 2016. The target is asteroid 101955 Bennu, with a rendezvous in 2018, and a sample return in 2023. The Asteroid Redirect Mission was eliminated from the 2018 budget. This was to capture an asteroid into Earth for study and mining.

The S-class asteroids have almost no water, but consist of metals such as iron, cobalt, and nickel. Also, there are trace elements in small fractions, but large numbers. A 10 meter S-type asteroid would be made up of about 650,000 kg of metal, of which 50 kg would be rare metals of great value. And the best part is, we don't need to bring it down to Earth to extract the metals. We do it in space, using solar power. Some of the material could be used in space construction.

Asteroid Amum 3554 is interesting to the Seattle-based company Planetary Resources. The M-class asteroid is about 1.6 km in size. It is estimated to contain a large amount of platinum, some $8 trillion ($10^{12}$) at current rates. Remember the Klondike gold rush? It also has massive amounts of iron, nickel, and cobalt. The project and company are well-financed by several entrepreneurial Billionaires. We know the composition by spectral analysis. Near Earth asteroid 1986DA about a mile wide, is estimated to contain 100,000 tons of platinum, and a mere 10,000 tons of gold.

We don't even need to go to the asteroid belt – there are at least 9,000 known near-Earth asteroid (NEO). It is suspected the count is low by several orders of magnitude. The potential retrieval and processing could be done by robotic missions.

Technically, an NEO is a solar system object whose closest approach to the Sun is 1.3 AU, and that comes in close proximity to the Earth There are 14,000 known asteroids in this category, 100 comets, solar orbiting spacecraft, and meteoroids. All these have the potential of striking the Earth. They are closely tracked from the ground, by NASA's Planetary Defense Coordination Office. A joint US/EU project called Spaceguard is tracking NEO's larger than 30 meters. Three NEO's have been visited by spacecraft.

Eros is the second largest Near Earth Object (NEO), 16.8 km in diameter. It was first seen in 1898. The NASA spacecraft NEAR-Shoemaker landed on Eros in 2001. This somewhat insignificant chunk of rock may be setting precedents in space law. It's ownership is claimed by an individual, Gregory W. Nemitz, who plans to exploit its resources, in violation of current Space Law. He also sent NASA a bill for the parking spot of the

Shoemaker-Levy lander. After quite a bit of court proceedings, specifically, the United States Court of Appeals for the Ninth District, the case has gone to a jury trail. The outcome will set precedent and define the ownership of space objects, and the exploitation of their resources.

At the moment, it seems the courts will agree with defined space Law.

The issue was addressed by the United Nations by the Outer Space Treaty of 1967, ratified by 102 countries. It makes countries responsible for the space activities of their citizens. The International Moon Treaty came into effect in 1984, and forbids private ownership of space resources or objects. Fifteen nations have ratified this treaty. There will be loopholes exploited, and nations can withdraw from the Outer Space Treaty with one years' notice. There will be a fertile and lucrative place for space lawyers for some time to come. We better figure this out quickly before we find a solid gold asteroid.

Mars Base

When we send the first human mission to Mars, it would be nice if we could refuel there for the return trip. There may be some water ice below the surface. It is also possible that Mars' tiny moons contain water ice. Why water ice? You use solar power to break it down into hydrogen and oxygen – the perfect rocket fuel. A big advantage is that we don't need to carry the return mission fuel with us on the outgoing flight. Another good fuel that may exist on Mars is methane.

The Players

This section discusses the companies who are active in developing approaches for manufacturing in space, as of this writing. In the haste to get this technology up and working, there are issues to be resolve, such as who owns what, who can exploit what, and who can be blamed. Similar to the Law of the Sea, there is a United Nations Law of Space, that most Nations have signed up to. It essentially says, you can't claim a celestial body. It's a bit vague on whether you can mine it for profit. Individuals and Companies are subject to the jurisdiction of their home country. Will this lead to small "safe-haven" countrys luring big investment by hosting space mining companies to operate under their venue? Think, off-shore banking, Flag of Convenience. Small countries, with large and wealthy backers don't necessarily abide with International norms. Might we need an International Space manufacturing organization, modeled on the International Maritime Organization? If so, it's getting late to get it organized. It's about to become another "gold rush." There is not, for example, a legal definition of where space begins.

Some of the existing organizations that address parts of the issue include the Inter-Agency Space Debris Coordination Committee; the Fourth Committee of the United Nations General Assembly, the Special Political and Decolonization committee; the United Nations Committee on the Peaceful Uses of Outer Space; and the United Nations Office for Space Affairs. None of these entities specifically address the ownership and usage of materials in space. There is an analogy to the exploitation of deep undersea mining. This is addressed in the U.N. Conventions on the Law of the Sea. There is an International Seabed Authority that has jurisdiction outside of a Nation's 200 mile Exclusive Economic Zone. Inside the zone, activities are regulated by the country's laws. There is also an International Institute of Space Law.

In addition, there are a number of relevant NGO's including the Commercial Spaceflight Federation, the European, Middle-East, and Africa Satellite Operators Association, and the Satellite Industry Organization.

As of this writing, there is no law regarding extracted natural resources from space. Sovereignty over any real property above the Earth's atmosphere is not currently defined.

"...the governmental policy toward the private or commercial space sector will have a significant impact on the business chances of those private space ventures." Secure World Foundation.

Commercial Spaceflight Federation

The CSF is a U.S. trade group for private spaceflight in general. It supports in-space manufacturing as one of its interest areas. Most of the players in this field are represented. It was founded in 2005, and currently has more than 70 members, including academic members. Besides promoting their title field, that intend to help the industry share best practices, and standards. The also provide fellowships to Academia. They help their members navigate the regulatory fields, and serve as a key point of contact with regulatory agencies. They also have a focus area on spaceports. They are located in Washington, D. C. As of late 2018, they have19 Executive members and 46 Associate members, as well as 21 Research and Education Affiliates.

Axion Space

Axion, of Houston, Texas, envisions a commercial replacement for the ISS. This will be used for research, manufacturing, and space tourism as early as 2020. Before their station is ready, they plan to deliver tourists to the ISS, on a non-interference basis to the station's mission. Their modular station will be built in an incremental manner. They will focus on tourism and in-space manufacturing as money-makers.

Axion's early effort will take the form of modules attached to the ISS, which will be removed and flown separately when the ISS is decommissioned. Manufacturing will be focused on 3-D printing, a proven space technology. They mention the market for jet turbines and the use of specialized 3D printing with metals done on Earth, which suffers from the gravity problem. They see the market for materials manufactured in space to be used in space-built projects, not necessarily just returned to Earth.

Orbital Technologies Commercial Space Station

This project is a 2010 proposed commercial effort in orbit by Orbital Technologies, a Russian company. The facility would be compatible with the Soyuz and Progress crew delivery and logistics craft. The project has support from the Russian Federal Space Agency, but no private funding as of this date.

Bigelow

Bigelow Aerospace is an American space company, headquartered in Las Vegas, NV. It was founded in 1998, and focuses on inflatable modular units, carried to orbit. The company licensed the technology from NASA, and has three Space Act agreements in place. The company has enhanced the technology.

Several modules have been sent to orbit, including Genesis-I (2006) and Genesis-II (2007). Both remain in orbit but are retired. Both have an expected life of 12 years in orbit, at which point they will reenter the atmosphere and burn. Both Genesis modules were heavily instrumented. Questions remain about their longevity and safety in the space environment.

The Bigelow Expandable Activity Module (BEAM) is in orbit, attached to the ISS. It was funded by NASA, and was launched on a SpaceX cargo mission. The emphasis of the inflatable modules is proving the radiation protection and debris shielding of the inflatable. It was launched in 2016, and will be evaluated through 2018.

Bigelow hopes to put one of their Expandable Bigelow Advanced Station Enhancement (XBASE) modules on the moon by 2021. This is referred to as the B330 Module. Space Complex Alpha is their next-generation commercial space station. Bigelow has launch agreements in place with SpaceX and Lockheed Martin, and has launched on the Russian Dneper vehicle.

Bigelow himself came out of the hospitality industry, and hopes to set up space hotels and habitats to encourage space tourism. One of these was the 2005 concept, CSS (Commercial Space Station) *Skywalker*. He is also addressing manufacturing in space, and has said he has agreements in place with six nations to pursue this.

His other concepts include the Next-Generation Commercial Space Station, and Space Complex Alpha, consisting of dual *Sundancer* Modules and a BA330. The BA330 *Nautilus* module has an enclosed volume of 330 cubic meters. The *Sundancer* would have had a volume of 180 cubit meters, but the project was canceled. The constraining factor is the lack of crew transportation systems, consisting, at the moment, of the Soviet Soyuz.

Excalibur-Almaz

Excalibur-Almaz is a private Russian company that has purchased two partially completed Almaz space station hulls.

Almaz "Diamond"was a Russian (Soviet) Military Reconnaissance station, identified as Salyut 2, 3, and 5. The program ran from 1973-1976. These Soyuz craft, were referred to as Orbital Pilot Stations (OPS). This program was similar to the USAF's cancelled MOL Project. The Almaz carried a 23mm automatic cannon, for self-defense. It was tested in orbit against a satellite target.

The company estimates they saved $2 billion dollars in development costs. These they plan to use for Space Tourism, and possibly for manufacturing in space. The space cannon were probably not included in the deal. The company is headquartered on the Isle of Man and successfully participated in NASA's Commercial Crew Development Program. The current status of the company is unknown, but in 2016, their equipment was converted into an education exhibit.

Space Island Group

The Space Island Group (www.spaceislandgroup.com) is addressing the manufacture of new materials in space. One product line would be thousands of new metal alloys, that are impossible to manufacture in a gravity field, This has been demonstrated on the International Space Station. The company plans to use a series of orbital modules, that can be leased, by corporations for their own projects.

Secure World Foundation

From their mission statement, "The Secure World Foundation envisions the secure, sustainable, and peaceful uses of outer space contributing to global stability on Earth. We work with governments, industry, international organizations, and civil society to develop and promote ideas and actions for international collaborations that achieve the secure, sustainable, and peaceful uses of outer space."

Goldman Sachs

This investment house has studied the asteroid mining scenario, and has reassured it clients about the favorable financial benefits of investing in this area. Investment houses are very conservative, so this is a favorable event for the emerging industry

Wrap-up

You can manufacture new materials in the zero-gravity of space that are not possible on Earth. It is incredibly expensive to transfer the raw material to orbit, manufacture it, and return it to Earth. Build it in space for use in space. More importantly, use locally obtained materials, obtained from the moon or asteroids. This is where specialty manufacturing is going. Don't get in its way.

It will be a wild frontier shortly, as commercial entities see the opportunity of making huge amounts of money (after spending huge amounts of money) from space-based resources. All this has to be carefully thought through and regulated. As with all historical projects on the frontier, these projects will drive regulation, as well as profits.

Bibliography

AIAA, *Space Manufacturing Eight: Energy and Materials from Space, Proceedings of the 10th Conference on Space Manufacturing,* 1992, ISBN-1563470225.

Aldrin, Buzz *No Dream Is Too High, Life Lessons From a Man Who Walked on the Moon*, National Geographic, 2016, ISBN-9781426216497.

American Welding Society, *Welding in Space and the Construction of Space Vehicles,* 1991, ISBN-9993984825.

Belfiore, Michael *Rocketeers, How a Visionary Band of Business Leaders, Engineers, and Pilots is Boldly Privatizing Space*, Harper Collins, 2008, ISBN-0061149039.

Bentley, Matthew A. *Spaceplanes: From Airport to Spaceport,* 2009, Springer, ASIN-B008BB7HQA.

Berinstein, Paula, Terenzi, Dr. Fiorella *Making Space Happen: Private Space Ventures and the Visionaries Behind Them*, 2002, ASIN-B00719IM12.

Berton, Pierre *The Klondike Fever: The Life And Death Of The Last Great Gold Rush*, 2015, ASIN-B06XGD1TCX.

Brynjolfsson, Erik, McAfee, Andrew *The Second*

Machine Age: Work, Progress, and Prosperity in a Time of Brilliant Technologies, 2014, ISBN-0393350649.

Dula, Arthur, Zhenjun, Zhang S*pace Mineral Resources: A Global Assessment of the Challenges and Opportunities*, 2015, ASIN-B018OJD95Q.

Edwards, Bradley C. and Westling, Eric A. *The Space Elevator: A Revolutionary Earth-to-Space Transportation System,* 2003, ISBN-0974651710.

Engler, Matthias "In-space Manufacturing Race," July 2017, avail: http://www.space-of-innovation.com/in-space-manufacturing-race/

Fassbender, Melissa "Drug Development in space: how microgravity enables pharma," 2016, avail: http://www.outsourcing-pharma.com/Preclinical-Research/Eli-Lilly-drug-development-in-space.

Grey, Jerry *Space Manufacturing Facilities (Space Colonies)*, 1977, AIAA, ASIN-B000VODUG6.

Gump, David *Space Enterprise Beyond NASA,* 1990, ASIN-B001J4ZK0Q.

Gurtuna, Ozgur *Fundamentals of Space Business and Economics*, 2013, SpringerBriefs in Space Development, ISBN-1461466954.

Harris, Phillip *Space Enterprise: Living and Working*

Offworld in the 21st Century, 2009, ASIN-B00DZ0PDP0.

Hudgins, Edward L *Space: The Free-Market Frontier,* 2003, Cato Institute, ASIN-B005HITTR0.

Krukin, Jeff, *NewSpace Nation: America's Emerging Entrepreneurial Space Industry, 2002,* 2nd Edition, ASIN-B00719IM12.

Ley, Willy, Rockets, *The Future of Travel Beyond the Stratosphere*, 1945, Viking Press, ASIN- 0007E7IC2.

Lewis, John S. *Asteroid Mining 101: Wealth for the New Space Economy*, 2014, Deep Space Industries, ASIN-B01LP8JMNQ.

Lewis, John S. *Mining the Sky: Untold Riches From The Asteroids, Comets, And Planets,* 1997, Helix Books, ISBN-0201328194.

Lewis, J., Matthews, M.S., and Guerrieri, M.L., (Ed), *Resources of Near-Earth Space,* University of Arizona Press, 1993, ISBN-978-0-8165-1404-5.

Matloff, Greg, Bang, C. *Harvesting Space for a Greener Earth*, 2014, ISBN-1461494257.

O'Neill, Gerald K. AIAA, *Space-Based Manufacturing from Nonterrestrial Materials,* 1977, Progress in Astronautics and Aeronautics, ISBN-10-0915928213.

National Research Council and Division on Engineering

and Physical Sciences,*3D Printing in Space*, 2014, National Academies Press, ISBN-10-0309310083.

Radley, Charles; Pearson, Jerome *The Lunar Elevator: Bringing the Riches of the Moon Down to Earth*, Springer. 2018, ISBN-3319664867.

Ross, Alec *The Industries of the Future*, Simon & Schuster, 2016, ASIN-B00UDCNJYO.

Samuelson, Paul A. and Nordhaus, William D. *Economics*, 2009, 19th ed, ISBN-0073511293.

Schwab, Klaus *The Fourth Industrial Revolution*, 2017, Crown Business, ASIN-B01JEMROIU.

Schulte-Ladbeck, Dr. Regina E. *Basics of Spaceflight for Space Exploration, Space Commercialization, and Space Colonization,* 2016, ISBN-150252595X.

Seedhouse, Eric *Suborbital: Industry at the Edge of Space*, Springer Praxis Books, 2014, ISBN-3319034847.

Seedhouse, Erik *Spaceports Around the World, A Global Growth Industry*, 2017, 1st ed, Springer, ISBN-3319468456.

Simpson, Michael K., Weeden, Brian C. *Handbook for New Actors in Space,* 2017, Secure World Foundation, ISBN-0692851410.

Stakem, Patrick H. *Robots and Telerobots in Space Applications*, 2011, PRRB Publishing, ASIN B0057IMJRM.

Stakem, Patrick H. *Lonaconing Residency, Iron Technology & the Railroad,* 2008, 2nd ed, ISBN-1520286422.

Swan, Peter *Space Elevators: An Assessment of the Technological Feasibility and the Way Forward,* 2013, ISBN-2917761318.

Thorpe, Andrew M. *The Commercial Space Station: Methods and Markets,* 2007, ISBN-978-1434327604.

Vance, Ashlee *Elon Musk: Tesla, SpaceX, and the Quest for a Fantastic Future*, 2015, Ecco, ISBN-0062301233.

Young, Anthony *The Twenty-First Century Commercial Space Imperative* (SpringerBriefs in Space Development), 2015, Spring, ISBN-331918928X.

Werkheiser, Niki, *In-Space Manufacturing (ISM): Pioneering Space Exploration,* avail: https://ntrs.nasa.gov/archive/nasa/casi.ntrs.nasa.gov/20150016175.pdf

Resources

- http://www.spaceislandgroup.com/manufacturing.html

- Partnerships to Advance the Business of Space, Sep 3, 2014 by Subcommittee on Science and Space of the Committee on Commerce, Science, and Transportation United States Senate.

- http://www.commercialspaceflight.org/

- Secure World Foundation, *Handbook for New Actors in Space,* avail: https://swfound.org/handbook/

- Vectors website - http://vc.airvectors.net/idx_sci.html

- https://www.space.com/29052-3d-printed-in-space-nasas-unboxing-video.htm

- https://www.popsci.com/factories-in-space

- https://www.forbes.com/sites/alexknapp/2017/08/31/made-in-space-is-successfully-taking-manufacturing-into-the-stars/#759c70e17d8d

- http://spacenews.com/space-manufacturing-and-the-last-mile/

- https://humanizing.tech/the-coming-gold-rush-of-space-manufacturing-601d9c2dd8b6.

- http://www.spaceislandgroup.com/manufacturing.html

- hrrp://www.ssi.org

- http://www.nss.org/settlement/manufacturing/princeton.htm

- Role of A Space Station in Pharmaceutical Manufacturing, avail: https://arc.aiaa.org/doi/abs/10.2514/6.1983-3116

- https://www.biopharmadive.com/news/outer-space-the-final-frontier-of-biopharma-rd/408522/

- https://www.nasa.gov/mission_pages/station/research/news/space_spiders_live.html

- Space Act, 2015 – avail: https://www.congress.gov/bill/114th-congress/house-bill/2262

- www.Permanent.com

- https://www.oceaneering.com/space-systems/

- http://madeinspace.us/archinaut/

- www.commercialspaceflight.org

- https://newspace.spacefrontier.org/2018/

- Inter-Agency Space Debris Coordination Committee: www.iadc-online.org/

- International Institute of Space Law, www.iislweb.org

- Off-Earth manufacturing: using local resources to build a new home". www.esa.int

- factoriesinspace.com

- Wikipedia, various.

Glossary of terms

Achondrite – asteroid rich in platinum metals.
AMF – additive manufacturing facility, 3D printer.
Apogee – furthest point in the orbit from the Earth.
Aphelion – furthest point to the Sun.
Apolune – furthest point to the Moon.
APS-1 (India) Asteroid Prospecting Satellite-1
ASIME - Asteroid Science Intersections with In-Space Mine Engineering.
ASIN – Amazon Standard Inventory Number.
ASME – American Society of Mechanical Engineers.
Astrionics – electronics for space flight.
BEAM – Bigelow Expandable Activity Module – commercial inflatable space module.
BEO – beyond Earth orbit.
CATALYST - Lunar Cargo Transportation and Landing by Soft Touchdown.
CBM – common berthing mechanism.
CCSDS – Consultative Committee for Space Data Systems.
Chrondrites – asteroids rich in water.
Cislunar – beyond Earth's atmosphere to just beyond the moon's orbit.
CM – crew module.
CME – Coronal Mass Ejection, blast of energetic particles from the Sun.
CMP – co-manifested payload.
CNSA – China National space Administration.
Conops – concept of operations.
Copuos – United Nations Committee on the Peaceful

Uses of Outer Space.
CPS – Cyrogenic Propulsion Stage.
CRTBP – Circular Restricted Three-body Problem.
Chondrite – non-metalic meteorite that has not been melted.
CSA – Canadian Space Agency, Agence Spatiale Canadienne.
CSF – Cislunar Support Flight; Commercial Spaceflight. Federation
CSS – Commercial Space Station (Bigelow), also Russian (Orbital Technologies).
C&W – caution and warning.
Cygnus – Orbital-ATK automated cargo vehicle for ISS.
Cyrogenic – relating to very low temperatures.
DAM – damage avoidance maneuver.
DCM – docking cargo module.
Delta-V – change in velocity.
DoD – (U.S.) Department of Defense
DRG – Distant Retrograde Orbit.
DRM – design reference mission.
DSG – Deep Space Gateway.
DSH – deep space habitat.
DSN – (NASA) Deep Space Network.
DST – Deep Space Transport.
DTM – dynamic test model, for structural tests.
ECLSS – Environmental Control & Life Support system.
EDL – Entry, Descent, Landing.
EM-x Exploration Mission number-x.
EML – ElectroMagnetic Levitator.
Ephemeris – position information data set for orbiting bodies, 6 parameters plus time.

Epoch – a moment in time for orbital elements.
EPS – electrical power system
EROS – an asteroid.
ESA – European Space Agency.
ESOA – European, Middle East, and Africa Satellite Operators Association.
EU – European Union.
EUS – Exploration Upper Stage.
EVA – extra-vehicular activity.
Exomedicine – study and exploration of medicines in zero gravity environment of space.
FAA – (U.S.) Federal Aviation Administration.
FCC – (U.S.) Federal Communications Commission.
FSL - Fluid Science lab
GEO – Geostationary Earth Orbit
GNC – Guidance, Navigation, and Control.
Gravity well – a conceptual model of the gravity field near a mass.
GSFC – NASA Goddard Space Flight Center, Greenbelt, MD.
GTO – Geosynchronous Transfer orbit.
Halo Orbit – three dimension orbit near the any of the 5 Lagrange points of two bodys.
HEEO – highly eccentric Earth orbit.
HEO – high Earth orbit.
HGTA - Habitat Ground Test Module. (ISS).
HITL – Human in the loop.
HSIR – human systems integration requirements
IAF – International Astronautical Federation
IDSS – International Docking System Standard.
IGA - (ISS) InterGovernmental Agreement.

IISL – International Institute of Space Law
ISM – in-space manufacturing.
ISO – International Standards Organization
ISP – specific impulse. Measure of efficiency of rocket engine, units of seconds.
ISRO – Indian Space Research Organization
ISRU – in situ resource utilization
ISS – International Space Station
I&T – Integration and Test.
ITU – International Telecommunication Union.
JAXA – Japan Aerospace Exploration Agency.
KW – kilowatt.
ISRU – in site resource utilization.
ISF – Industrial Space Facility
ISS – International Space Station
JAXA – Japanese space agency
JPL – Jet Propulsion Laboratory, Pasadena, CA.
JSC – Johnson Space Center, Houston, Texas.
KSC – NASA Kennedy Space Center, launch site, Florida.
L2 – second of 5 Lagrange points, a null in the gravity field in the restricted 3-body problem.
LAS – launch abort system.
Lbf – pounds, force.
LCT – Lunar Cargo Transportation.
LEO – Low Earth Orbit
LH2 – liquid hydrogen.
Libration point – null in the gravity field of the three body problem.
LOS – (Russian) Lunar Orbital Station; loss-of-signal.
LOX – liquid oxygen, boils at -297 F.

LSAM – lunar surface access module.
LSPPO – Lunar Surface Systems Project Office (NASA-JSC).
LST – landing by soft touchdown.
MADV – Mars Ascent/Descent Vehicle.
MBC – Mars Base Camp.
MET – mission elapsed time.
Microgravity – almost zero gravity; Weightlessness.
MMSEV – MultiMission Space Exploration Vehicle.
MOL – (USAF) Manned orbiting lab project.
MOU – memorandum of understanding.
MPCV - Multi-Purpose Crew Vehicle.
MPLM – Multi-purpose Logistics Module.
m/s – meters per second.
MSG – microgravity science glovebox.
MSL – Material Science laboratory.
MSL-EMG - Material Science laboratory, Electromagnetic Levitator.
Mt – metric ton, 1000 kg.
NAC – NASA Advisory Council.
Nadir – the point directly below.
NASA – (U.S.) National Aeronautics and Space Administration.
NEO – near Earth object.
NextSTEP-2 – (NASA) Next Space Technologies of Exploration Partnerships.
NGO – non-governmental organization.
NHV – net habitable volume.
NOAA – (U. S.) National Oceanographic and Atmospheric Administration.
NRHO – Near rectilinear halo orbit (around the L1 or L2

Earth-Moon libration point).
NTIS – National Technical Information Service (www.ntis.gov).
NTRS – NASA Technical Reports Server, ntrs.nasa.gov.
OOSA – (U.N.) Office for Outer Space Affairs.
OPSEK – orbital piloted assembly and experiment complex.
ORU – Orbital Replacement Unit.
OST – outer space treaty.
Perigee –closest point in the orbit from the Earth.
Perhelion – closest point to the Sun.
Perilune – closest point to the Moon.
PMA – Pressurized mating adapter.
PMCU – Power Management Control Unit.
PPB – power and propulsion bus.
PMM - (ISS) Permanent Multipurpose Module
PTCS – Passive thermal control system.
PVCU – Photo Voltaic Control Unit.
RCS – reaction control system.
R&D – research & development.
Regolith – layer of loose material, covering rock; dirt.
ROSCOSMOS – Russian Space Agency.
RPOD – Rendezvous, Proximity Operations, Docking.
SEP – solar electric propulsion,
SHFE – space human factors engineering.
SI – System International – the metric system.
SIA – Satellite Industry Association,
Sidereal period – time for an object to make a full orbit.
Sol, local solar day – on Mars, 24h, 37 min., on the moon, 14 days.
SLS – (NASA) Space Launch System.

SPACE Act - Spurring Private Aerospace Competitiveness and Entrepreneurship.
SPR - Science Payload Rack (ISS)
STM – Space Traffic Management.
Synodic period - time for an object in orbit to occupy the same point, in relation to 2 other objects.
TCS – thermal control system.
TLI – Trans-lunar injection.
TM – Technical Manual.
TPS – thermal protection system.
Trillion - 10^{12}
TRL – technology readiness level.
UDM – universal docking module.
Ullage – residual fuel or oxidizer in a tank after engine burn is complete.
UN – United Nations.
USAF – United States Air Force.
V&V – verification and validation.
WDV – water delivery vehicle.
XBASE - Expandable Bigelow Advanced Station Enhancement.
Zenith – the point directly above.
Zombie-sat – a non functional satellite in orbit, contributing to the orbital debris problem.

If you enjoyed this book, you might also be interested in some of these.

Stakem, Patrick H. *16-bit Microprocessors, History and Architecture*, 2013 PRRB Publishing, ISBN-1520210922.

Stakem, Patrick H. *4- and 8-bit Microprocessors, Architecture and History*, 2013, PRRB Publishing, ISBN-152021572X,

Stakem, Patrick H. *Apollo's Computers*, 2014, PRRB Publishing, ISBN-1520215800.

Stakem, Patrick H. *The Architecture and Applications of the ARM Microprocessors*, 2013, PRRB Publishing, ISBN-1520215843.

Stakem, Patrick H. *Earth Rovers: for Exploration and Environmental Monitoring*, 2014, PRRB Publishing, ISBN-152021586X.

Stakem, Patrick H. *Embedded Computer Systems, Volume 1, Introduction and Architecture*, 2013, PRRB Publishing, ISBN-1520215959.

Stakem, Patrick H. *The History of Spacecraft Computers from the V-2 to the Space Station*, 2013, PRRB Publishing, ISBN-1520216181.

Stakem, Patrick H. *Floating Point Computation*, 2013, PRRB Publishing, ISBN-152021619X.

Stakem, Patrick H. *Architecture of Massively Parallel Microprocessor Systems*, 2011, PRRB Publishing, ISBN-1520250061.

Stakem, Patrick H. *Multicore Computer Architecture,* 2014, PRRB Publishing, ISBN-1520241372.

Stakem, Patrick H. *Personal Robots*, 2014, PRRB Publishing, ISBN-1520216254.

Stakem, Patrick H. *RISC Microprocessors, History and Overview,* 2013, PRRB Publishing, ISBN-1520216289.

Stakem, Patrick H. *Robots and Telerobots in Space Application*s, 2011, PRRB Publishing, ISBN-1520210361.

Stakem, Patrick H. *The Saturn Rocket and the Pegasus Missions, 1965,* 2013, PRRB Publishing, ISBN-1520209916.

Stakem, Patrick H. *Visiting the NASA Centers, and Locations of Historic Rockets & Spacecraft,* 2017, PRRB Publishing, ISBN-1549651205.

Stakem, Patrick H. *Microprocessors in Space*, 2011, PRRB Publishing, ISBN-1520216343.

Stakem, Patrick H. Computer *Virtualization and the Cloud*, 2013, PRRB Publishing, ISBN-152021636X.

Stakem, Patrick H. *What's the Worst That Could Happen? Bad Assumptions, Ignorance, Failures and Screw-ups in Engineering Projects, 2014,* PRRB Publishing, ISBN-1520207166.

Stakem, Patrick H. *Computer Architecture & Programming of the Intel x86 Family, 2013,* PRRB Publishing, ISBN-1520263724.

Stakem, Patrick H. *The Hardware and Software Architecture of the Transputer*, 2011,PRRB Publishing, ISBN-152020681X.

Stakem, Patrick H. *Mainframes, Computing on Big Iron*, 2015, PRRB Publishing, ISBN- 1520216459.

Stakem, Patrick H. *Spacecraft Control Centers*, 2015, PRRB Publishing, ISBN-1520200617.

Stakem, Patrick H. *Embedded in Space,* 2015, PRRB Publishing, ISBN-1520215916.

Stakem, Patrick H. *A Practitioner's Guide to RISC Microprocessor Architecture*, Wiley-Interscience, 1996, ISBN-0471130184.

Stakem, Patrick H. *Cubesat Engineering*, PRRB Publishing, 2017, ISBN-1520754019.

Stakem, Patrick H. *Cubesat Operations*, PRRB Publishing, 2017, ISBN-152076717X.

Stakem, Patrick H. *Interplanetary Cubesats*, PRRB Publishing, 2017, ISBN-1520766173 .

Stakem, Patrick H. Cubesat Constellations, Clusters, and Swarms, Stakem, PRRB Publishing, 2017, ISBN-1520767544.

Stakem, Patrick H. *Graphics Processing Units, an overview*, 2017, PRRB Publishing, ISBN-1520879695.

Stakem, Patrick H. *Intel Embedded and the Arduino-101, 2017,* PRRB Publishing, ISBN-1520879296.

Stakem, Patrick H. *Orbital Debris, the problem and the mitigation*, 2018, PRRB Publishing, ISBN-*1980466483*.

Stakem, Patrick H. *Manufacturing in Space*, 2018, PRRB Publishing, ISBN-1977076041.

Stakem, Patrick H. *NASA's Ships and Planes*, 2018, PRRB Publishing, ISBN-1977076823.

Stakem, Patrick H. *Space Tourism*, 2018, PRRB Publishing, ISBN-1977073506.

Stakem, Patrick H. *STEM – Data Storage and Communications*, 2018, PRRB Publishing, ISBN-

1977073115.

Stakem, Patrick H. *In-Space Robotic Repair and Servicing*, 2018, PRRB Publishing, ISBN-1980478236.

Stakem, Patrick H. *Introducing Weather in the pre-K to 12 Curricula, A Resource Guide for Educators*, 2017, PRRB Publishing, ISBN-1980638241.

Stakem, Patrick H. *Introducing Astronomy in the pre-K to 12 Curricula, A Resource Guide for Educators*, 2017, PRRB Publishing, ISBN-198104065X.
Also available in a Brazilian Portuguese edition, ISBN-1983106127.

Stakem, Patrick H. *Deep Space Gateways, the Moon and Beyond*, 2017, PRRB Publishing, ISBN-1973465701.

Stakem, Patrick H. *Exploration of the Gas Giants, Space Missions to Jupiter, Saturn, Uranus, and Neptune*, PRRB Publishing, 2018, ISBN-9781717814500.

Stakem, Patrick H. *Crewed Spacecraft*, 2017, PRRB Publishing, ISBN-1549992406.

Stakem, Patrick H. *Rocketplanes to Space*, 2017, PRRB Publishing, ISBN-1549992589.

Stakem, Patrick H. *Crewed Space Stations,* 2017, PRRB Publishing, ISBN-1549992228.

Stakem, Patrick H. *Enviro-bots for STEM: Using Robotics in the pre-K to 12 Curricula, A Resource Guide for Educators,* 2017, PRRB Publishing, ISBN-1549656619.

Stakem, Patrick H. *STEM-Sat, Using Cubesats in the pre-K to 12 Curricula, A Resource Guide for Educators,* 2017, ISBN-1549656376.

Stakem, Patrick H. *Lunar Orbital Platform-Gateway,* 2018, PRRB Publishing, ISBN-1980498628.

Stakem, Patrick H. *Embedded GPU's,* 2018, PRRB Publishing, ISBN- 1980476497.

Stakem, Patrick H. *Mobile Cloud Robotics,* 2018, PRRB Publishing, ISBN- 1980488088.

Stakem, Patrick H. *Extreme Environment Embedded Systems,* 2017, PRRB Publishing, ISBN-1520215967.

Stakem, Patrick H. *What's the Worst, Volume-2,* 2018, ISBN-1981005579.

Stakem, Patrick H., *Spaceports,* 2018, ISBN-1981022287.

Stakem, Patrick H., *Space Launch Vehicles,* 2018, ISBN-1983071773.

Stakem, Patrick H. *Mars,* 2018, ISBN-1983116902.

Stakem, Patrick H. *X-86, 40th Anniversary ed*, 2018, ISBN-1983189405.

Stakem, Patrick H. *Lunar Orbital Platform-Gateway*, 2018, PRRB Publishing, ISBN-1980498628.

Stakem, Patrick H. *Space Weather*, 2018, ISBN-1723904023.

Stakem, Patrick H. *STEM-Engineering Process*, 2017, ISBN-1983196517.

Stakem, Patrick H. *Space Telescopes,* 2018, PRRB Publishing, ISBN-1728728568.

Stakem, Patrick H. *Exoplanets*, 2018, PRRB Publishing, ISBN-9781731385055.

Stakem, Patrick H. *Planetary Defense*, 2018, PRRB Publishing, ISBN-9781731001207.

Patrick H. Stakem *Exploration of the Asteroid Belt*, 2018, PRRB Publishing, ISBN-1731049846.

Patrick H. Stakem *Terraforming*, 2018, PRRB Publishing, ISBN-1790308100.

Patrick H. Stakem, *Martian Railroad,* 2019, PRRB Publishing, ISBN-1794488243.

Patrick H. Stakem, *Exoplanets,* 2019, PRRB Publishing, ISBN-1731385056.

Patrick H. Stakem, *Exploiting the Moon,* 2019, PRRB Publishing, ISBN-1091057850.

Patrick H. Stakem, *RISC-V, an Open Source Solution for Space Flight Computers,* 2019, PRRB Publishing, ISBN-1796434388.

Patrick H. Stakem, *Arm in Space*, 2019, PRRB Publishing, ISBN-9781099789137.

Patrick H. Stakem, *Extraterrestrial Life*, 2019, PRRB Publishing, ISBN-978-1072072188.

Patrick H. Stakem, *Space Command*, 2019, PRRB Publishing, ISBN-978-1693005398.

CubeRovers, A Synergy of Technologys, 2020, PRRB Publishing, ISBN-979-8651773138.

Robotic Exploration of the Icy moons of the Gas Giants. 2020, PRRB Publishing, ISBN- 979-8621431006

Hacking Cubesats, 2020, PRRB Publishing, ISBN-979-8623458964.

History & Future of Cubesats, PRRB Publishing, ISBN-979-8649179386.

Hacking Cubesats, Cybersecurity in Space, 2020, PRRB Publishing, ISBN-979-8623458964.

Powerships, Powerbarges, Floating Wind Farms: electricity when and where you need it, 2021, PRRB Publishing, ISBN-979-8716199477.

Hospital Ships, Trains, and Aircraft, 2020, PRRB Publishing, ISBN-979-8642944349.

<u>2020/2021 Releases</u>

CubeRovers, a Synergy of Technologys, 2020, ISBN-979-8651773138

Exploration of Lunar & Martian Lava Tubes by Cube-X, ISBN-979-8621435325.

Robotic Exploration of the Icy moons of the Gas Giants, ISBN- 979-8621431006.

History & Future of Cubesats, ISBN-978-1986536356.

Robotic Exploration of the Icy Moons of the Ice Giants, by Swarms of Cubesats, ISBN-979-8621431006.

Swarm Robotics, ISBN-979-8534505948.

Introduction to Electric Power Systems, ISBN-979-8519208727.

Centros de Control: Operaciones en Satélites del Estándar CubeSat (Spanish Edition), 2021, ISBN-979-8510113068.

Exploration of Venus, 2022, ISBN-979-8484416110.

Patrick H. Stakem, *The Search for Extraterrestial Life,* 2019, PRRB Publishing, ISBN-1072072181.

The Artemis Missions, Return to the Moon, and on to Mars, 2021, ISBN-979-8490532361.

James Webb Space Telescope. A New Era in Astronomy, 2021, ISBN-979-8773857969.

www.ingramcontent.com/pod-product-compliance
Lightning Source LLC
Chambersburg PA
CBHW030507220526
45464CB00006B/2693